FLYING

Curriculum Consultants

Dr. Arnold L. Willems
Associate Professor of Curriculum and Instruction
The University of Wyoming

Dr. Gerald W. Thompson
Associate Professor
Social Studies Education
Old Dominion University

Dr. Dale Rice
Associate Professor
Department of Elementary and Early Childhood Education
University of South Alabama

Dr. Fred Finley
Assistant Professor of Science Education
University of Wisconsin

Subject Area Consultants

Astronomy
Robert Burnham
Associate Editor
Astronomy Magazine and *Odyssey* Magazine

Geology
Dr. Norman P. Lasca
Professor of Geology
University of Wisconsin — Milwaukee

Oceanography
William MacLeish
Editor
Oceanus Magazine

Paleontology
Linda West
Dinosaur National Monument
Jensen, Utah

Physiology
Kirk Hogan, M.D.
Madison, Wisconsin

Sociology/Anthropology
Dr. Arnold Willems
Associate Professor of Curriculum and Instruction
College of Education
University of Wyoming

Technology
Dr. Robert T. Balmer
Professor of Mechanical Engineering
University of Wisconsin — Milwaukee

Transportation
James A. Knowles
Division of Transportation
Smithsonian Institution

Irving Birnbaum
Air and Space Museum
Smithsonian Institution

Donald Berkebile
Division of Transportation
Smithsonian Institution

Zoology
Dr. Carroll R. Norden
Professor of Zoology
University of Wisconsin — Milwaukee

First Steck-Vaughn Edition 1992

First published in Great Britain by Macmillan Children's
Books, a division of Macmillan Publishers Ltd, under the
title *Look It Up*.
First edition copyright © 1979, 1981 Macmillan Publishers Ltd
(for volumes 1-10)
First edition copyright © 1980, 1981 Macmillan Publishers Ltd
(for volumes 11-16)
Second edition copyright © 1985, 1986 Macmillan Publishers Ltd

Text this edition copyright © 1986 Raintree Publishers Inc., a
Division of Steck-Vaughn Company.

Library of Congress Number: 86-567

  3 4 5 6 7 8 9 0   97 96 95 94 93 92

Library of Congress Cataloging-in-Publication Data

Let's discover flying.

  (Let's discover; 14)
  Bibliography: p. 69
  Includes index.
  Summary: A reference book dealing with balloons, gliders
and all manner of aircraft, with special section on flying
animals and on airports.
  1. Aeronautics—Juvenile literature. 2. Airplanes—Juvenile
literature [1. Aeronautics. 2. Airplanes]
I. Title: Flying. II. Series.
AG6.L43 vol 14, 1986 [TL547] 031s [629.13] 86-567
ISBN 0-8172-2613-3 (lib. bdg.)
ISBN 0-8172-2594-3 (softcover)

LET'S DISCOVER
FLYING

97-316

RSVP
RAINTREE
STECK-VAUGHN
PUBLISHERS
The Steck-Vaughn Company

Austin, Texas

Contents

FLYING ANIMALS

Many animals can move through the air, but only birds, bats, and insects can fly. They flap their wings to stay in the air. Other animals, like the flying lizard, only glide. The flying fish swims fast, then leaps out of the water. Spiders are carried through the air on a silk thread.

gull

flying fish

butterfly

bat

flying lizard

spider

How birds fly

Birds are well designed for flying. Their feathers help to lift them and move them in the air. Powerful breast muscles help them flap their wings. Different birds have different shaped wings. Swifts have narrow wings for flying fast. Vultures have broad wings.

vultures

hummingbird

Hummingbirds get their name from the noise their wings make. They beat their wings about 80 times a second. They are the only birds that can fly backwards.

A bird's wings do two things. They lift the bird upward. They also push the bird forward.

1

8

On the upstroke the bird glides along freely. Its wings are twisted so that it is not pushed down and back again.

On the downstroke the wings are spread out. They push down and back. This pushes the bird forward and up.

9

Flying insects

Most insects have two sets of wings. They beat their wings very fast. A mosquito beats its wings more than 500 times in a second.

The bumblebee gets its name from the buzzing sound made by its fast wings.

bumblebee

longhorn beetle

Beetles have two pairs of wings. The front pair fit over its back to protect the delicate back wings. These hard, shiny lids are spread out to uncover the back wings, which are used for flying.

The housefly has only one pair
of wings. Behind the wings are a
pair of tiny clubs. The clubs help
the fly to balance.

club

Gliding animals

Some animals can glide long distances. They do not have proper wings. Flying squirrels and phalangers have loose flaps of skin between their legs. The flying frog has large webbed feet. The flying snake can flatten its body.

phalanger

phalanger

flying snake

flying squirrel

flying frog

13

FLYING MACHINES

For thousands of years people wished they could fly like birds. There are many stories about people who tried to fly.

Today we still cannot fly like birds. But aircraft can carry people quickly and safely through the air.

An old Greek story tells how Icarus used wings of wax and feathers to fly with. The sun melted the wax, and he fell into the sea.

One very old story is about King Bladud of England. The story says he made a pair of wings for himself. He wanted to fly over London. When he tried the wings out, he could not fly. He fell to the ground and was killed.

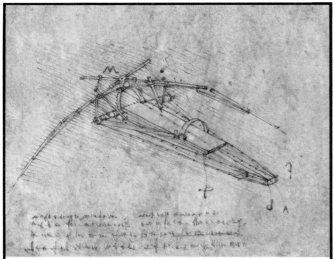

Leonardo da Vinci lived 500 years ago. He made drawings of machines with flapping wings. This was his idea for a flying machine.

Kites were made in China 5,000 years ago. It was not until 1852 that a kite big enough to lift a person was made. In this picture, a man is being lifted by a big kite.

The first aircraft looked like big kites. They had two sets of wings and no engines. They were called gliders. They were fastened to a rope and pulled along by cars or boats.

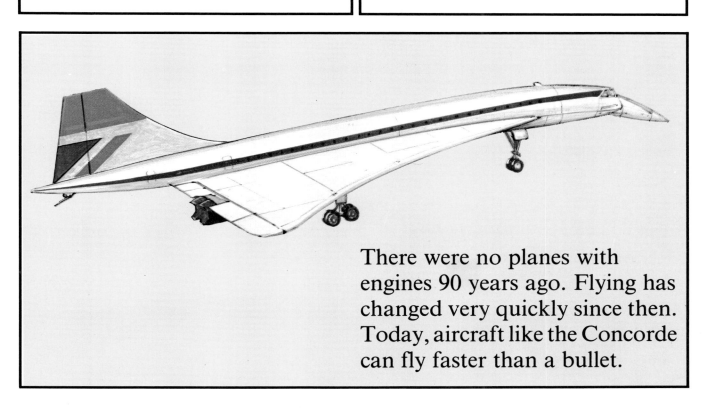

There were no planes with engines 90 years ago. Flying has changed very quickly since then. Today, aircraft like the Concorde can fly faster than a bullet.

Balloons

People flew in balloons long before airplanes were made. A Frenchman, Jacques Charles, flew a balloon in 1783. It was filled with a gas called hydrogen. Hydrogen is lighter than air. It made the balloon rise into the air. Balloons cannot be steered. They are blown along by the wind.

Most balloons today are filled with hot air. Ballooning is a popular sport.

In 1783, two Frenchmen, the Montgolfier brothers, were the first people to ride in a hot air balloon. Fire heated the air inside the balloon. As the air got hot, the balloon rose into the sky.

Gliders

A glider is an aircraft that has no engine. It is usually fastened to a small plane by a strong cable. The plane takes off and tows the glider behind it. Once in the air, the cable is dropped away and the glider floats free.

Gliding is an exciting sport. This glider can be taken apart and put on a trailer. It can be taken home after a flight.

Sir George Cayley made a glider in 1852. It carried one person and had a tailplane and a rudder to steer with. A copy of this glider was made a few years ago. You can see it above being towed by a car. It flew very well.

The Wright brothers were the first people to build an airplane with an engine that worked. At first they made gliders like the one below.

19

Airships

An airship is a balloon driven by an engine. The first one was invented by Henri Giffard in 1852. Later airships could carry up to 50 passengers. Some flew across the Atlantic. The trip was quiet but slow. Some airships broke up in the air or caught fire.

This airship is fastened to a mast to keep it from floating away. Passengers went up the mast in an elevator to get on board.

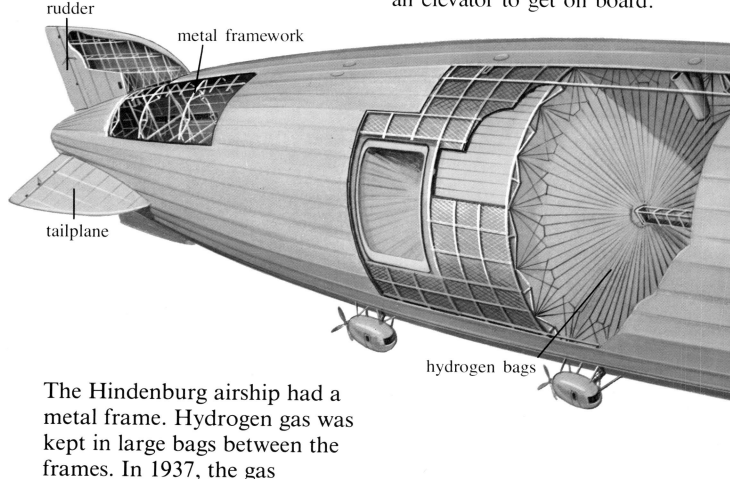

rudder

metal framework

tailplane

hydrogen bags

The Hindenburg airship had a metal frame. Hydrogen gas was kept in large bags between the frames. In 1937, the gas exploded, destroying the ship.

blimp

This airship is a blimp. It does not have a metal frame. The gas inside gives it its shape.

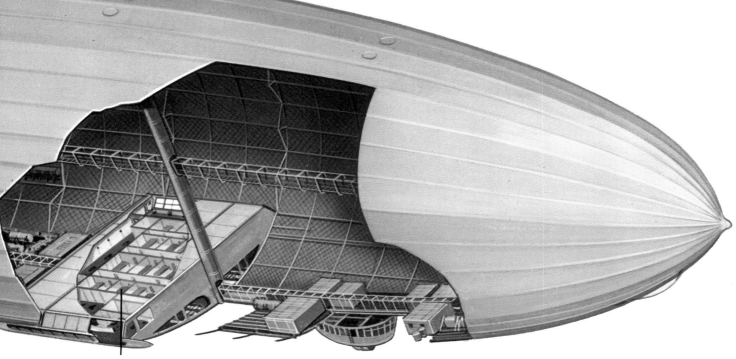

passenger area

AIRCRAFT

Early aircraft

Orville and Wilbur Wright were bicycle makers. They became interested in flying. First they built kites and gliders. This taught them about flight. They learned how the air moves. They also learned to make light, strong aircraft.

First the Wright brothers made kites. Strings were tied to the wings. A person on the ground pulled the strings to keep the wings level.

This glider kite was large enough to carry Wilbur Wright. He flew in it for 10 seconds. He found that aircraft controls must work very quickly.

The first flight
December 17, 1903

The Wright brothers were clever builders. After four years of trying, they built the first real plane. They learned how to control it and turn it in the air. They learned how to land safely. They were the first people to fly in a heavier-than-air machine.

3

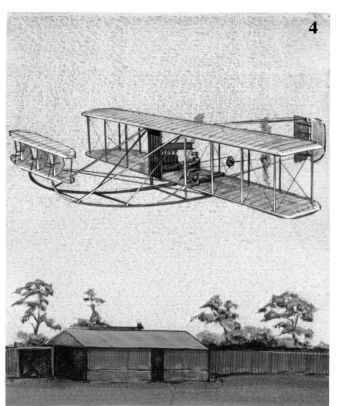

4

In 1903 Orville flew their first real plane. It had a gas engine and two propellers. It stayed in the air for 12 seconds and traveled about 37 meters (120 feet).

The Wright brothers kept working. Soon their plane could fly for as long as an hour. It had seats for the pilot and a passenger.

Early sporting aircraft

In 1908 the Wright brothers took one of their planes to Europe. More people became interested in building and flying aircraft. There were many new designs. Flying soon became a popular sport. Races were held. People went to flying shows to see the new aircraft.

Blériot

Farman

Farman

Farman

Antoinette

97-316

Planes for war and peace

Special fighter planes were built in the First World War. They carried guns. They could shoot at soldiers on the ground or other aircraft in the sky. Later in the war, bomber planes were made. After the war, some companies that made bombers made the first airliners.

Handley Page airliner

This was one of the first war planes. The propeller was behind the pilot's seat. A second person sat in front and fired the gun.

Vickers FB5 biplane

This German fighter plane had three wings. The famous pilot the Red Baron flew in one of these planes.

Fokker triplane

The record breakers

After the war, new planes with more powerful engines were built. Aircraft could fly farther without refueling.

Pilots flew to many parts of the world. Soon, airlines were started. They carried people to distant countries.

In 1927 Charles Lindbergh flew alone from New York to Paris. It took him 33 hours.

In 1919 Alcock and Brown made the first nonstop crossing of the Atlantic. It took them 16 hours. Brown had to knock ice off the plane's wings.

Amy Johnson flew from London to Australia in 1930. The trip took 19 days.

In 1931 the first planeload of airmail arrived in England from Australia in this plane.

Seaplanes

Seaplanes and flying boats take off and land on the water. This large craft is a flying boat. It was built in 1938 to carry heavily loaded seaplanes into the air.

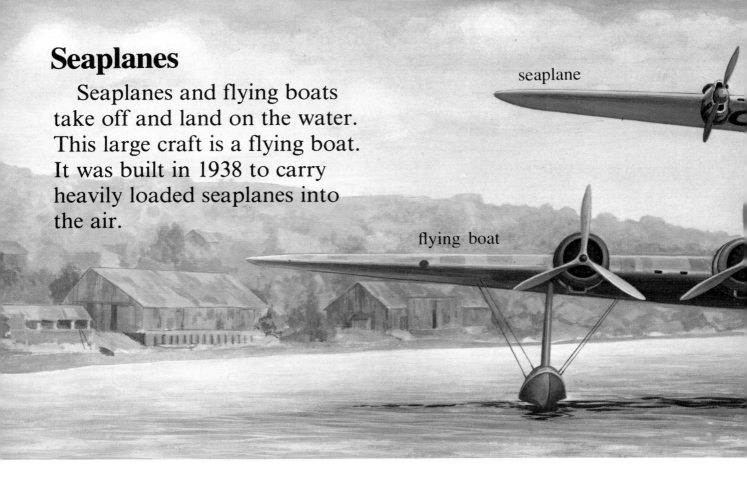

seaplane

flying boat

Seaplanes like the one above are sometimes used for sport. They cannot fly very fast. The floats slow them down. On the right is a large flying boat for passengers.

float

ANTILLES AIR BOATS
VIRGIN ISLANDS

VP-LVE

This is a Japanese amphibian plane. It can land on water or a runway. This one is used to rescue people at sea. It can take off and land in rough seas.

Modern aircraft

Here are some famous aircraft. You can see how much planes have changed. Most of the early planes were biplanes. They had two wings. One-winged planes are called monoplanes. They flew faster and were stronger. Later, jets were invented. They fly very fast.

This monoplane flew from France to England in 1909.

This big Russian biplane, built in 1913, had 4 engines.

This German monoplane was one of the first metal aircraft.

This airliner flew people from England to Africa and India.

This American plane of 1935 was the first modern airliner.

6

This British flying boat flew to Africa and Australia.

7

This American plane was the first big aircraft to fly high.

8

The British Comet was the first jet airliner.

9

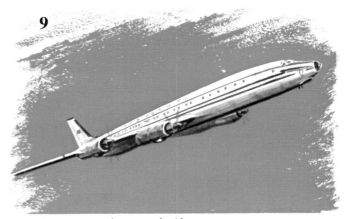

This Russian airliner was once the fastest propeller plane.

10

This huge plane was one of the first wide-bodied jets.

11

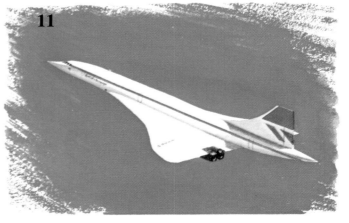

The first faster-than-sound airliner was the Concorde.

Boeing 747

Passenger planes

Over 500 million people travel by jet airliner every year. The Boeing 747 jumbo jet is the largest airliner. It can carry about 500 people. Airliners with propellers may be used for short trips. The Dash 7 carries 70 people. It has very quiet engines.

Most big passenger planes are made in the United States. The only big one made in Europe is the Airbus. Flying used to be very expensive. Large planes have made air travel much cheaper for everyone.

Early airliners were slow, noisy, and cold. The bumpy ride often made passengers sick.

Aircraft at war

Many of the planes made today are war planes. Some are bombers. Others are fighter planes that can shoot down other aircraft.

This American F-15 fighter flies at more than twice the speed of sound.

The British Tornado is a jet with swing wings. When the wings are spread out, the jet can fly slowly. It can fly twice as fast as the speed of sound when the wings are folded back.

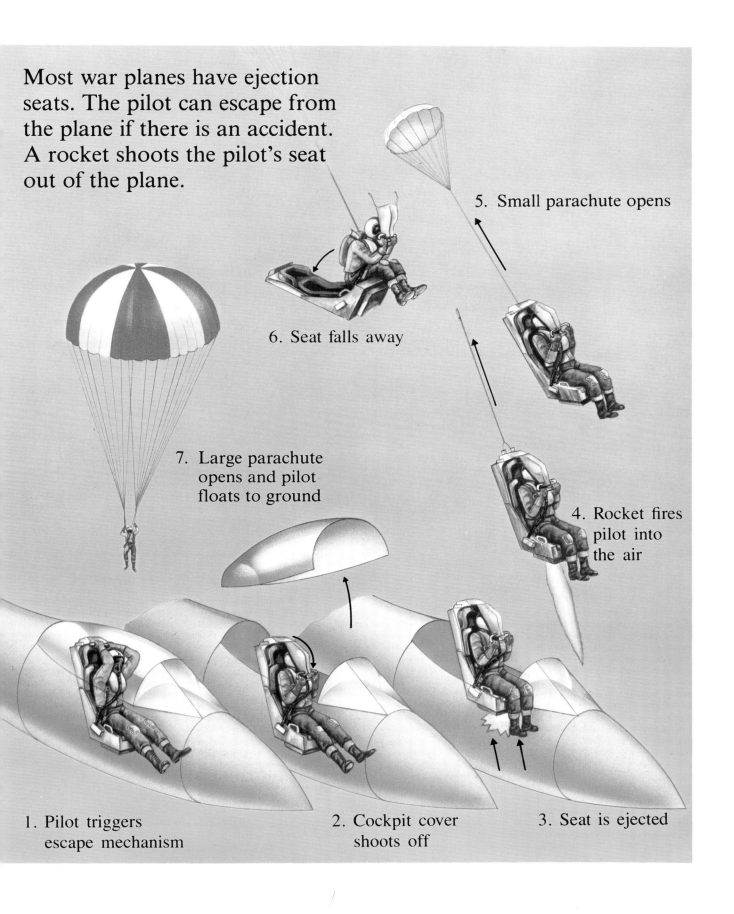

Most war planes have ejection seats. The pilot can escape from the plane if there is an accident. A rocket shoots the pilot's seat out of the plane.

5. Small parachute opens

6. Seat falls away

7. Large parachute opens and pilot floats to ground

4. Rocket fires pilot into the air

1. Pilot triggers escape mechanism

2. Cockpit cover shoots off

3. Seat is ejected

Supersonic planes

Sound travels at 1190 kilometers (739 miles) per hour. Supersonic planes can travel faster than sound. These planes have pointed bodies and swept-back wings. Supersonic aircraft are noisy. They make a loud bang when they reach the speed of sound.

This Russian MIG has a sharp pointed shape. Its shape helps it to fly faster than sound.

This transport plane is not supersonic. Its rounded nose and thick wings slow it down.

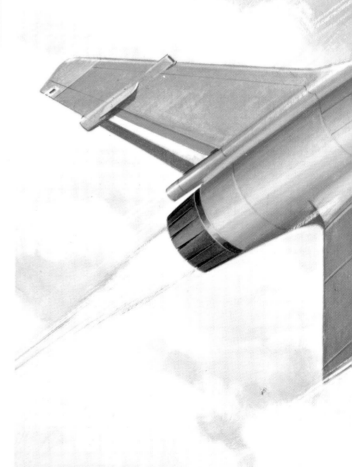

Many supersonic planes have wings shaped like a triangle. This is called a delta wing. This French Mirage jet has a delta wing. It can fly very fast.

The Concorde and the Space Shuttle

The first supersonic aircraft were powered by rockets. These aircraft were used to learn about flying at high speeds. Engineers built fast fighter planes and bombers. Then they made plans for a supersonic passenger airliner.

The Concorde has a narrow delta wing. It flies smoothly at 2177 kilometers (1,350 miles) per hour. It can fly much higher than other airliners. Here you can see the many people who are needed to get the plane ready for flight.

At the left, the Space Shuttle is being carried on top of a jumbo jet for testing. The Space Shuttle is part plane and part spacecraft. For regular use, rockets shoot it into space. It can return to earth and land like an ordinary airplane.

Helicopters

Helicopters do not need runways. They take off and land straight up and down. The large rotor on top keeps the helicopter in the air. The small rotor on the tail is used to steer it. Helicopters can hover in the air. They are useful for rescuing people.

This helicopter is landing on a snowy mountain in Greenland.

Helicopters are very useful to the armed forces. Large helicopters can carry guns and tanks to wherever they are needed. They can be used to move troops quickly.

This is a helicopter crane. It is taking a heavy load to a building site in the mountains.

43

Cargo aircraft

Some planes carry goods rather than people. They are called cargo planes. They have wide doors to load and unload crates and containers. Sometimes they carry special loads such as racing cars or zoo animals. Even elephants can travel by plane!

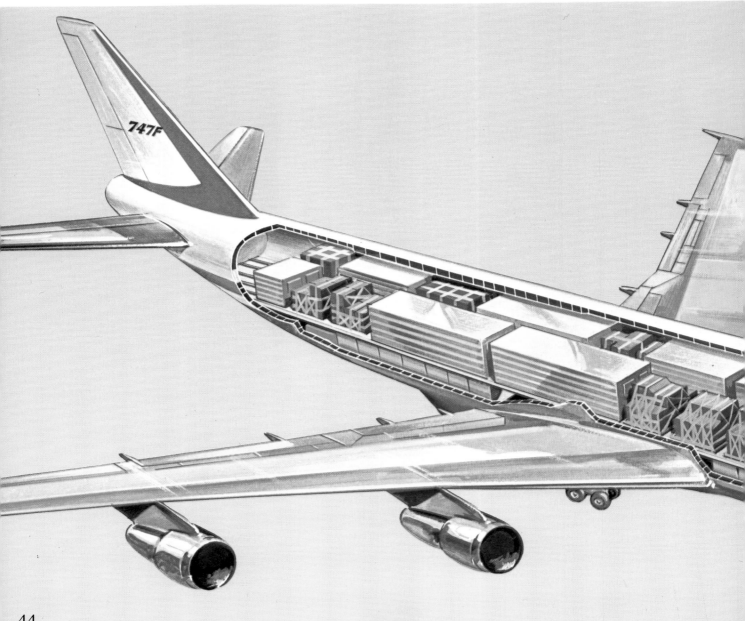

This jumbo cargo plane has a nose that opens for loading and unloading. The plane above has a tail that swings open.

The Super Guppy, above and below, carries parts of a plane called the Airbus A300. It takes them to a factory where they are put together. It is an odd shape so that it can carry large parts.

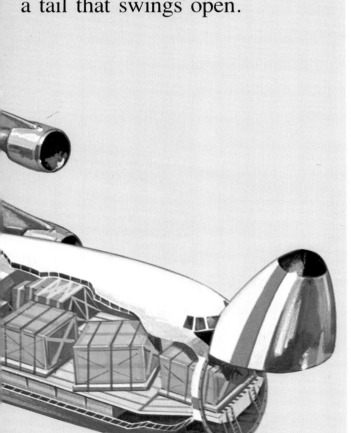

Some planes have been made into special car freighters. People going on a vacation can take their cars with them. The cars travel in one part of the plane, and the passengers sit in another part.

Small aircraft

Not all modern planes are big and fast. Here you can see some small aircraft. Some look like the early planes, but they are safer and more powerful. They have special instruments to help the pilot navigate at night and in bad weather. Some people own their own planes. They use them for pleasure or for business.

This is the Pitts Special biplane. A skilled pilot can do many tricks in the air with it.

The Cessna factory makes more than 8,000 of these small planes every year.

This light plane can take off and land in a very short space. The wings are up above its body and it is called a high-wing monoplane.

This smaller size jet carries 10 passengers. It can fly almost as fast as a big airliner.

Special aircraft

Small planes have many uses. In some places, doctors reach sick people by plane. Planes spray farm crops to keep them healthy. People make maps by taking pictures from aircraft. Planes are used to study the weather. These children are having a geography lesson.

Airplanes are used to fight forest fires. They drop water or chemicals on the fires. Fire fighters parachute into a forest to help put out a fire.

HOW AIRCRAFT ARE MADE

Here are some of the inventions used in modern planes. This plane can take off and land like a helicopter. Air from the engines is pumped out of nozzles. When the nozzles point down, the plane rises up off the ground. When the nozzles are turned backwards, the plane flies forward.

Jet engines suck in air through the intakes. The air is mixed with fuel and burned in the engine. This makes a very hot gas. Gas rushing out of the engine nozzles pushes the plane forward.

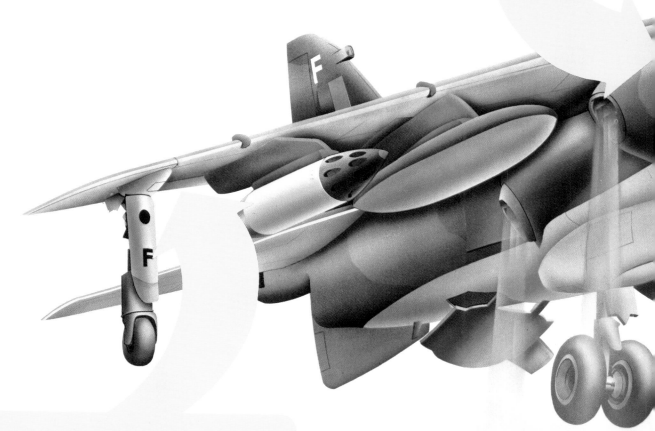

Ailerons are hinged flaps on the wings. The pilot uses them to keep the plane level or to make turns.

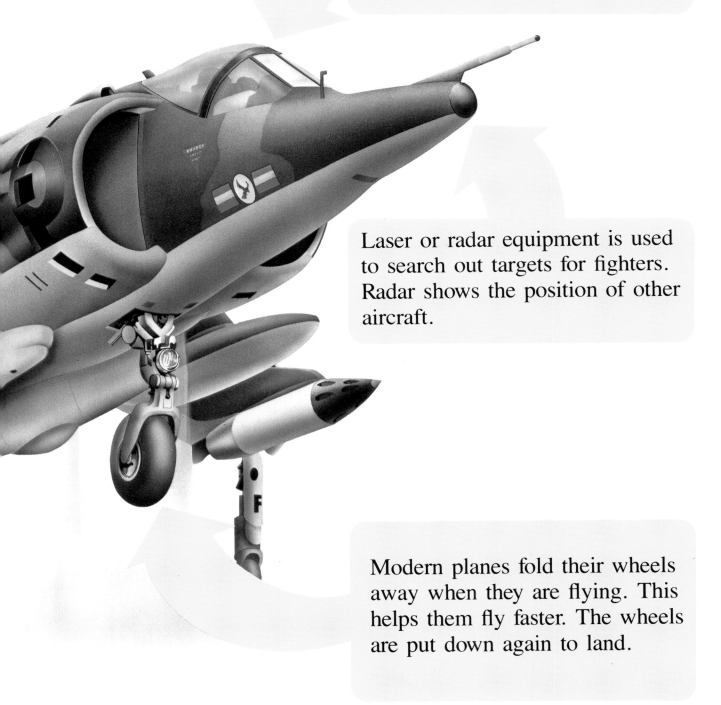

A pilot has over 300 controls and indicators to watch. All the automatic devices are controlled from the cockpit.

Laser or radar equipment is used to search out targets for fighters. Radar shows the position of other aircraft.

Modern planes fold their wheels away when they are flying. This helps them fly faster. The wheels are put down again to land.

Making planes

A new aircraft is designed on paper. A model is tested in a wind tunnel. Changes are made until the design is right.

Plans are drawn of all the different parts needed. It takes many people and a long time to make a new plane.

model in wind tunnel

drafter designing aircraft

This is an aircraft factory.
Thousands of parts are put
together to make a plane.

AIRPORTS

Airports are very busy places. They cover a huge area. Millions of passengers pass through a large airport every year. At a busy airport, a plane lands or takes off every minute. How many planes can you see in this picture?

Pilots and crew

This is a drawing of a Boeing 747 jumbo jet. It can carry up to 550 passengers and has a crew of 17. Three of the crew sit in the cockpit and control the plane. They are the captain, the first officer, and the flight engineer. The rest of the crew are flight attendants.

The pilot of an airliner is called the captain. The captain flies the plane.

The first officer is the copilot. The copilot sits next to the captain and helps him.

A flight engineer sits at the back of the cockpit and makes sure everything works properly.

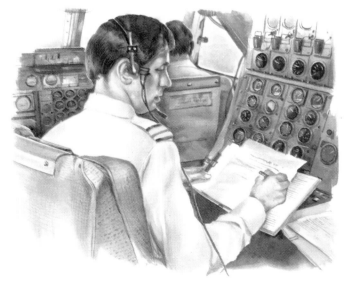

The Boeing 747 can fly more than 15,000 kilometers (9,300 miles) without stopping to refuel.

Flight attendants serve food and drink to the passengers and help keep them comfortable.

Flight attendants work very hard. They look after the passengers. They are trained to help in all kinds of emergencies.

vet

nurse

waiter

airport
police

baggage
handler

telephone
operator

fire
fighter

ground
attendant

customs
officer

engineer

mechanic

cleaner

cook

Airport staff

Many people are needed to get an airliner ready for flight. The ground staff work at the airport. Some fill the plane with fuel. Some see that the plane is working properly. Others handle baggage or make food for the passengers.

Air traffic control

Air traffic controllers tell pilots when to land and take off. Radar helps them to know the position of all the planes flying to and from the airport. Flight paths are mapped out through the air. Aircraft are guided along the path by radio. Controllers tell pilots where the other aircraft are. This helps to avoid midair collisions.

Planes of the future

People are still working to make better and faster aircraft. They are trying all kinds of new designs. Some of these new ideas are shown here. Early fliers would be amazed to see how much planes have changed in the last 80 years.

Plans are being drawn up for a larger version of the Concorde. It may look like this.

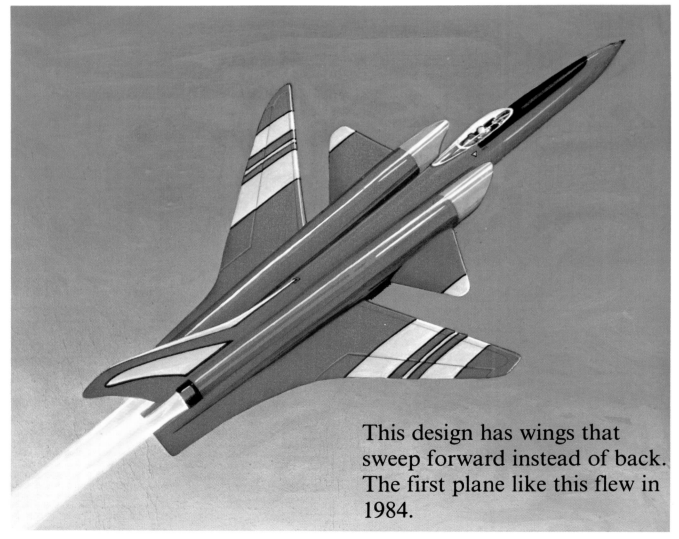

This design has wings that sweep forward instead of back. The first plane like this flew in 1984.

The Space Shuttle is already a success. Some day there will be many space planes.

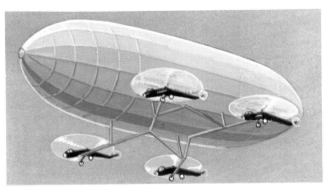

This is a combination of an airship and a helicopter. Models of it are being tested.

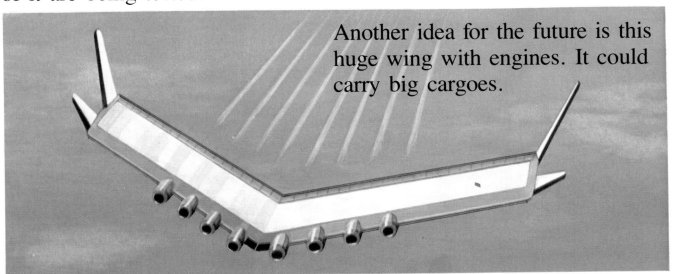

Another idea for the future is this huge wing with engines. It could carry big cargoes.

GLOSSARY

These words are defined the way thay are used in the book.

aileron (AY luh RAHN) the hinged flaps on the wings of an airplane that keep it level

Airbus A300 (EHR buhs) a jet that has been built by many countries. It can land in a short space

aircraft (EHR kraft) any machine made to fly in the air

airliner (EHR ly nuhr) an airplane that carries passengers from one place to another

airline (EHR lyn) a company that owns many airplanes and flies passengers to distant places

airmail (EHR mayl) mail carried between cities by airplane

airport (EHR pawrt) a place with fields for airplanes to land and take off

airship (EHR shihp) an aircraft that is lighter than air and is driven by an engine

air traffic controller (ehr TRAF ihk kuhn TROHL uhr) the person who tells the pilot when to land and take off and makes sure planes do not crash into each other

amphibian plane (am FIHB ee uhn playn) an aircraft that can land on either water or a runway

aviation (AY vee AY shuhn) the art of flying aircraft

automatic device (aw tuh MAT ihk dih VYS) something that operates by itself

baggage (BAG ihj) the suitcases, trunks, and bags that a person takes when traveling

balloon (buh LOON) a bag filled with air or another gas

beetle (BEET'l) a flying insect with hard shiny front wings

biplane (BY playn) an aircraft with two sets of wings

blimp (blihmp) an airship that does not have a metal frame

Boeing (BOH ihng) the name of an airliner

bomber (BAHM uhr) an aircraft used to drop bombs

bumblebee (BUHM buhl bee) a flying insect that makes a noise with its wings

cabin (KA bihn) the part of the plane where the passengers sit

captain (KAP tuhn) the person who flies and controls an aircraft; another name for the pilot

cargo (KAHR goh) another name for freight

Cessna (SEHS nuh) a company that builds small planes

chemical (KEHM ih kuhl) a

substance made by or used in chemistry; gases and acids are chemicals

cockpit (KAHK piht) the place where the pilot sits and controls the airplane

commercial (kuh MUR shuhl) relating to business or trade; the commercial airliner flew people as a business

Concorde (KAHN cawrd) a large supersonic jet airplane

controls (kuhn TROHLS) the instruments for guiding a machine

copilot (KOH py luht) the person who helps the captain; another name for the first officer

crane (krayn) a large machine that can move heavy objects from one place to another

crew (kroo) a group of people who work together to make something run

delta wing (DEHL tuh wihng) a three-sided wing

elevator (EHL uh VAY tuhr) a small room or cage that can be raised or lowered and is used to move people or goods

ejector seat (ee JEHK tahr seet) a special seat that shoots out of a plane so that the pilot can escape quickly if there is an accident

engine (EHN jihn) a machine that uses energy to run other machines

factory (FAK tuhr ee) a place where things are built

feather (FEHTH uhr) one of the light growths that cover a bird's skin

fighter plane (FY tuhr playn) an aircraft that carries guns

first officer (fuhrst AW fuh suhr) the person who helps the captain and sits in the cockpit

flight attendant (FLYT uh tehn duhnt) the person who takes care of the passengers by serving them food and drink

flight engineer (flyt EHN juh NEER) the person who checks the indicators and makes sure the plane is working properly

flight path (flyt path) the line or way along which an airplane moves

float (floht) to rest on top of water or a liquid; floats are objects that help an airplane sit on water

flying boat (FLY ihng boht) a large aircraft that takes off and lands on water

flying squirrel (FLY ihng SKWUR uhl) a small, furry animal with flaps of skin between its legs that help it glide

geography (jee AHG ruh fee) the study of the surface of the earth

glide (glyd) to move smoothly along without any effort

glider (GLY duhr) an aircraft that has no engine

helicopter (HEHL uh KAHP tuhr) an aircraft that is kept in the air by blades that rotate above it

high-wing monoplane (hy wihng MAH noh playn) an airplane with one wing above its body

hover (HUV uhr) to stay in the air, flying right above one place

hummingbird (HUHM ihng BURD) a tiny bird that makes a humming noise with its wings; the only bird that can fly backwards

hydrogen (HY druh juhn) a gas that is lighter than air

indicators (IHN dih kay tohrs) the dials that tell the pilot how the airplane is running

insect (IHN sehkt) an animal with three body parts and six legs; most insects have two sets of wings

instrument (IHN struh muhnt) a tool used for doing a certain kind of work

intake (IHN tayk) a place in a channel or pipe where a liquid or gas is taken in

jet (jeht) an airplane with a jet engine

jet engine (jeht EHN jihn) a machine that is driven by a stream of liquid, gas, or vapor forced through a small opening

jumbo jet (JUHM boh jeht) the name of the largest jet aircraft

kite (kyt) a light wooden frame covered with paper, plastic, or cloth that flies on the end of a string

laser (LAY zuhr) a machine that makes a very strong beam of light

Lindbergh, Charles (LIHND burg chahr'lz) the first pilot that flew from New York to Paris in 33 hours

mast (mast) a tall metal cage that an airship is fastened to when not flying

monoplane (MAH noh playn) an aircraft with one set of wings

mosquito (muhs KEE toh) a flying insect that sucks blood when it bites

nozzle (NAHZ uhl) a spout at the end of a hose or pipe

parachute (PAR uh SHOOT) a large device made of fabric that is shaped like an umbrella; it is used to drop people or things safely to the ground from an airplane

passenger (PAS uhn juhr) a person who travels

phalanger (fuh LAN juhr) a small, furry animal with flaps of skin between its legs that help it glide

pilot (PY luht) the person who flies and controls an aircraft

propeller (pruh PEHL uhr) a twisted blade mounted at an angle around a hub; when a propeller turns, it moves air or water and helps drive an airplane or boat forward

radar (RAY dar) a machine that uses radio waves to find and track objects

Red Baron (rehd BEHR uhn) a famous German pilot from World War I

refuel (ree FYOO uhl) to load up with more fuel

rescue (REHS kyoo) to save or free

rocket (RAHK iht) a device that is driven forward by a stream of hot gases that are released from the rear

rotor (ROH tuhr) the part of a motor that turns or rotates

rudder (RUHD uhr) the part of an airplane's tail that allows the plane to turn

runway (RUHN way) a long, narrow area where an airplane can take off and land

seaplane (SEE playn) an aircraft that takes off and lands on water

Space Shuttle (spays SHUT uhl) an aircraft that is part plane and part spacecraft. It takes people into space

steer (steer) to guide the course of; the pilot steered the plane onto the runway

swift (swihft) a bird with narrow wings for flying fast

swing-wing (SWIHNG wihng) wings that have two positions; forward and back

Super Guppy (SOO puhr GUHP ee) a large cargo plane that carries parts for the Airbus A300

supersonic (SOO puhr SAHN ihk) something that flies faster than sound

tailplane (TAYL playn) the back part of an airplane

Tornado (tawr NAY doh) the name of a British swing-wing jet airplane

transport (TRANS pawrt) to carry freight or people from one place to another

triangle (TRY an guhl) an object with three sides and three angles

vulture (VUHL chuhr) a large bird with broad wings that it uses for gliding

war plane (wawr playn) an aircraft used in war; usually a bomber or a fighter

wind tunnel (wihnd TUHN uhl) a passage through which air is blown at a known speed; it is used in experiments

wing (wihng) a body part that a bird, insect, or other animal uses to fly; also, part of an airplane

Wright, Orville and Wilbur (ryt, AWR vihl and WIHL buhr) two brothers who were the first people to fly in a heavier-than-air machine

FURTHER READING

Allen, John E. *Early Aircraft*. New York: Silver Burdett Company, 1979.

Burchard, Peter. *Balloons: From Paperbags to Skyhooks*. New York: Macmillan, 1960. 48pp.

Colby, C. B. *Jets of the World: New Fighters, Bombers and Transports*. New York: Coward, 1966. 48pp.

Corbett, Scott. *What Makes A Plane Fly?* Boston: Little, Brown and Company, 1967. 58pp.

Davidson, Jesse. *Famous Firsts in Aviation*. New York: Putnam, 1974.

Delear, Frank J. *Famous First Flights Across the Atlantic*. New York: Dodd, Mead and Company, 1979.

Dolan, Edward F. *Great Mysteries of the Air*. New York: Dodd, Mead, 1983.

Dwiggins, Don. *Flying and the Frontiers of Space*. New York: Dodd, Mead, 1982.

Dwiggins, Don. *Why Airplanes Fly*. Chicago: Childrens Press, 1976. 31pp.

Feravolo, Rocco V. *Junior Science Book of Flying*. Champaign, Illinois: Garrard, 1960. 65pp.

Foster, Genevieve. *Year of the Flying Machine*. New York: Scribner, 1977.

Foster, John T. *The Flight of the Lone Eagle: Charles Lindbergh Flies Nonstop from New York to Paris*. New York: F. Watts, Inc., 1974. 61pp.

Freedman, Russell. *How Birds Fly*. New York: Holiday House, 1977. 64pp.

Gilleo, Alma. *Air Travel From the Beginning*. Elgin, Illinois: Child's World, Inc., 1977.

Graves, Charles P. *The Wright Brothers*. new
ed. New York: Putnam, 1973.

Harris, Susan. *Helicopters*. New York: F. Watts,
Inc., 1979. 48pp.

Hewish, Mark. *Know Your Aircraft*. Chicago:
Rand McNally, 1977.

Holland, Sharon. *Wings to Rockets*.
Windermere, Florida: Rourke, 1982.

Hunt, L. C., ed. *Adventure of Flight*. New
York: Holt, Rinehart and Winston, 1973.

Kanetzke, Howard W. *Airplanes and Balloons*.
Milwaukee: Raintree Childrens Books, 1978.
33pp.

Kershaw, Andrew. *Airlines*. Windermere,
Florida: Rourke, 1981.

Moran, Tom. *Kite Flying is for Me*.
Minneapolis: Lerner Publications, 1984.

Navarro, John G. *Superplanes*. Garden City,
New York: Doubleday, 1979. 79pp.

Peterson, David. *Airports*. Chicago:
Childrens Press, 1981.

Shepherd, Walter. *How Airplanes Fly*. New
York: John Day Company, 1972.

Stein, R. Conrad. *The Story of the Flight at
Kitty Hawk*. Chicago: Childrens Press,
1981.

Tunney, Christopher. *Aircraft*. Minneapolis:
Lerner Publications, 1980.

Weston, Graham. *In the Air*. New York:
Bookwright Press, 1983.

Wilson, Mike and Robin Scagell. *Jet Journey*.
New York: Viking Press, 1978. 59pp.

QUESTIONS
TO THINK ABOUT

Flying Animals

Do you remember?

What groups of animals can really fly?

What are some animals that can glide through the air and seem to fly?

How do feathers help birds to fly?

What is unusual about the hummingbird?

How do the wings of swifts and vultures differ?

How many wings do most insects have?

What does a beetle use its front wings for?

What helps the flying frog travel through the air?

Find out about . . .

Butterflies. What are butterflies? What do they look like when they are born? How do they change? How many kinds of butterflies are there?

Birds. What are some of the kinds of birds that live in your part of the country? Where do they nest? How do they take care of their young? How do young birds learn to fly?

Flying Machines

Do you remember?

Who was Icarus? What happened to him?

What are two things that were used to make balloons lift into the sky?

What did the Montgolfier brothers use to fly?

How does a glider work?

How does an airship differ from a balloon?

Why did people stop using airships after a while?

Find out about . . .

Icarus. Who was Icarus? Who made his wings of feathers and wax? Find out more about this old Greek story, or myth.

Early ballooning. When did ballooning begin? In what country or countries? Who were some of the early people who made and flew in balloons?

Gliders. How are gliders used today? How have they been used in war? What are modern gliders like? Where can people fly in gliders?

Aircraft

Do you remember?

Why are the Wright brothers important?

When did the Wright brothers make their first flight in a real airplane?

How long did their first flight last? How far did they travel?

Where did the pilot sit in the Vickers FB5

warplane? Who sat in front of him?

What did the Red Baron's plane look like?

What record-breaking trip did Alcock and
Brown make? In what year?

Who was the first person to fly alone from New
York to Paris?

Name a famous early woman pilot. What
record-breaking trip did she make?

What is the difference between a sea plane and
a flying boat?

Where can an amphibian plane take off and
land?

What is a biplane? What is a monoplane?

What country built the first jet airliner? The
fastest propeller plane?

What plane is the largest jet airliner? How many
passengers can it carry?

What is an ejector seat? When is it used?

What is a supersonic plane?

How fast can the Concorde fly?

What is a delta wing?

What is unusual about the Space Shuttle?

What does the big rotor on a helicopter do?
What does the small one do?

What does the Super Guppy carry? Why does it
have such a strange shape?

How do fire fighters use airplanes?

Find out about . . .

The Wright brothers. Who were the Wright brothers? Where did they live? When? How did they get interested in flying? What were some of the planes they built?

Early airplanes. What did the early airplanes look like? Who made them? What countries built early airplanes? How were early planes used? What was it like to fly them?

Record breakers. Who were some of the record breakers in flying? When did they make their record-breaking flights? What problems did they have?

Amelia Earhart. Who was Amelia Earhart? What flights did she make? What happened to her?

Seaplanes. What do modern seaplanes look like? What are they used for?

Amphibian planes. What does an amphibian plane look like? Where are such planes used? What are they used for? Who makes them?

Jumbo jets. What are some of the biggest planes made? How many passengers can they carry? How fast can they travel. How far can they go without stopping to refuel?

Fighter planes. What are some of the fastest fighter planes made today? How fast are they? What arms do they carry?

74

Supersonic planes. What is the speed of sound? What plane first flew faster than sound? How fast do the fastest planes travel today? What is a sonic boom?

The Concorde. Who made the Concorde? Where is it used? How fast is it? How many passengers does it carry?

The Space Shuttle. Why was the Space Shuttle built? How can it be used? How does it work?

Helicopters. Who invented the helicopter? When? How does a helicopter work? What are helicopters used for today? What are the biggest helicopters?

How Aircraft Are Made

Do you remember?

Where are a plane's ailerons? What are they used for?

How does a jet engine work?

What is radar used for?

What is a wind tunnel used for?

Find out about . . .

Radar. What is radar? How does it work? When was radar invented? How was it first used? What is it used for today?

Making planes. What are some of the companies in the United States that make planes? Where are their factories? What are some different kinds of jobs in an airplane factory? What kinds of training are needed to do these jobs?

Airports

Do you remember?

How many people pass through a busy airport in a year?

How often do planes take off and land at a busy airport?

How many people are in the crew of a Boeing 747?

What does the captain of an airliner do? What do flight attendants do?

Who are some of the people who work on an airport staff?

What do air traffic controllers do?

Find out about . . .

Airports. What is the busiest airport in the United States? How many people pass through it? How often do planes take off or land? What are some special problems airports can have?

Airport jobs. What are some different jobs that need to be done at an airport? What kinds of training do people need to do these jobs? Who can be a flight attendant?

PROJECTS

Project — A Book of Early Fliers

Make a book of your own about some of the early fliers. First, decide which fliers you want to include. It can be some of the people you read about in this book. You can also include others you know about.

Use one sheet of paper for each chapter of your book. Use both sides. Write a short story about one or two fliers on each sheet. Include some pictures. You can draw them yourself or paste in pictures from other places, like old magazines.

Include pages for the title of your book and for the contents. Number the pages. Finally, use staples or a piece of string through holes on the left side of the sheets to bind your pages together. You may want to add stiff paper covers as well.

Project — Take a Trip

Take a trip to learn more about airplanes and flying. Here are two places you might visit.

1. A local airport. Here you can watch planes take off and land. Some airports have guided tours.

2. A museum. Some museums have models of early aircraft. Some even have real aircraft that were used by early fliers.

INDEX

Photo Credits: Airbus Industrie; Anglia TV; Ardea; Peter J. Bish; Austin J. Brown; Philip Clark; Bruce Coleman; Graham Finch; Imperial War Museum; Matt Irvine; Radio Times Hulton Picture Library; Spectrum Colour Library; ZEFA.

Front cover: Experimental Aircraft Assn./Joan Seybold.

Illustrations: Jim Bamber; Robert Burns; Dick Eastland; Philip Emms; Colin Hawkins; Ron Haywood; Eric Jewell; Vanessa Luff; Wilfred Plowman; Mike Roffe; George Thompson; Raymond Turvey; Michael Whelply.